T0173136

interesting
INSECTS

Gavin Broad, Blanca Huertas,
Ashley Kirk-Spriggs & Dmitry Telnov

First published by the Natural History Museum, Cromwell Road, London SW7 5BD
© The Trustees of the Natural History Museum, London 2020
Reprinted with updates 2022

All rights reserved. No part of this publication may be transmitted in any form or by any means without prior permission from the British Publisher

The Authors have asserted their right to be identified as the Authors of this work under the Copyright, Designs and Patents Act 1988

ISBN 978 0 565 09503 1

A catalogue record for this book is available from the British Library

10 9 8 7 6 5 4 3 2

Reproduction by Saxon Digital Services, Norfolk, UK
Printed by Toppan Leefung Printing Limited, China

Introduction

Insects first inhabited the land over 400 million years ago and have flourished ever since. A factor to their overwhelming success is their evolutionary ability to conquer all the Earth's environments, except marine. Approximately 80 per cent of all animal species are insects, and there are thought to be anywhere between a whopping 10 million and 100 million species alive today. This huge number of species is grouped into 26 to 30 orders, based on major differences in body plans and mode of metamorphosis. The 'big four' insect orders that dominate are beetles (Coleoptera); bees, wasps and their relatives (Hymenoptera); flies (Diptera); and butterflies and moths (Lepidoptera). All of these incredibly species-rich orders have complete metamorphosis, from an egg to a larva to a pupa to an adult.

Of the groups with incomplete metamorphosis where the young, nymphs, don't look radically different from the adults, the true bugs (Hemiptera) are especially species-rich, and the crickets, grasshoppers and locusts (Orthoptera), and dragonflies and damselflies (Odonata), are very visible. Examples of all but the dragonflies and damselflies are included in this book. From a basic body plan of six legs, antennae, and an articulated exoskeleton, insects have become pollinators, predators, waste disposers, parasites, pests and carriers of disease.

The insects in this book show a diverse array of colour, patterns and form – insects really do come in all shapes and sizes – and they range from micro wasps to enormous birdwing butterflies. All specimens featured are from London's Natural History Museum collection and measurements are approximate for the insect species as a whole. The authors hope you are as amazed and fascinated by the selection as they are.

Red and black froghopper

Cercopis vulnerata

The red and black froghopper is found on woody or herbaceous plants, mainly in wooded areas. It readily takes to the wing when disturbed and is also equipped with very effective jumping back legs, allowing leaps of up to 70 cm (2¼ ft). The nymphs are rarely seen, as they feed on underground roots.

Distribution
Temperate Europe

Size
10–12 mm (approximately ½ in) long

Six-spot burnet

Zygaena filipendulae

Feeding on bird's-foot trefoil, the six-spot burnet is a familiar sight in meadows. The moths are active by day, with nothing much to fear from predators as they are full of toxic hydrogen cyanide. The caterpillars are black and yellow and have a similar chemical defence. However, their cocoons are heavily parasitized by a wasp, *Listrognathus obnoxius*, which prefers larger female caterpillars, meaning some six-spot burnet populations end up with many more males.

Distribution
Europe
Size
35–40 mm (1½–1½ in) wingspan

Jewel beetle

Temognatha chalcodera

Larvae of this brightly coloured jewel beetle are woodborers and their reproductive biology is unusual. Eggs are laid in clutches on timber or bark and covered by a presumably protective coating of material, until the larva hatches and bores into the wood. Even more unexpected is that the female uses fine sand or charcoal powder from burnt timber to case the eggs, and her ovipositor has special receptacles to collect and store this material.

Distribution
Western Australia

Size
43–50 mm (1¾–2 in) long

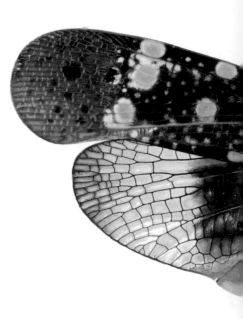

Mottled fulgorid plant bug

Desudaba danae

This fulgorid bug (from the family Fulgoridae) is confined to a few coastal rainforest fragments of Queensland, Australia. The fore wings are dark brown to black, with rows of yellow spots, while the membranous hind wings are red and brown at the base. Fulgorid bugs have sucking mouthparts and feed on tree sap.

Distribution
Australia

Size
10–12 mm (approximately ½ in) long

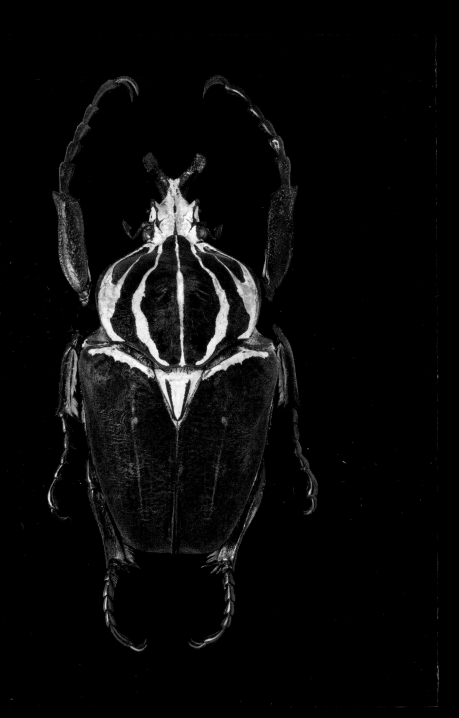

Goliath beetle

Goliathus goliatus

Being among the largest and heaviest insects on Earth, Goliath beetles occur in tropical rainforests of Africa. Males have a short Y-shaped horn on the head and are normally brown and white or black and white. Females are usually smaller and range from a dark brown to silky white. Each leg ends in a pair of strong sharp claws that are useful for climbing on tree trunks and branches.

Distribution
Equatorial Africa

Size
50–110 mm (2–4¼ in) long

Seven-spot ladybird

Coccinella septempunctata

Ladybirds are among the most popular insects - they widely appear as toys, in animation and children's books. Ladybirds secrete a fluid from joints in their legs, which gives them a foul taste. A threatened ladybird may also play dead very well. Looking so cute, these beetles and their funny-looking larvae are in fact ferocious predators of aphids and other small herbaceous insects, protecting our gardens and crops effectively and providing this service for free.

Distribution
Europe, Asia, northern Africa, introduced to Australia, tropical Africa and North America

Size
7–11 mm (¼–½ in) long

Red spotted jewel beetle

Stigmodera cancellata

This beauty is endemic to coastal Western Australia where the larvae live in soil and feed on the roots of myrtle shrubs for up to 15 years! Adults emerge with perfect timing to coincide with the wildflower season – October to November. Hardened fore wings (elytra) are coarsely punctured, greenish or bluish with six irregular red spots and red lateral margins, while the forebody is green, coppery or blackish. The females are significantly larger than the males.

Distribution
Western Australia
Size
23–35 mm (1–1½ in) long

Orange-spotted fruit chafer

Amaurodes passerinii

The elytra, hardened fore wings, of this majestic fruit chafer are dark with rows of circular red to orange spots, while the thoracic shield (pronotum) can be either pale, bicoloured or dark. The complete upper surface of the beetle is covered in dense short velvety hairs. Only males show forward-projecting horns – a reliable weapon in their fierce fights for females. The adults feed on tree sap or nectar and are active in sunny hot weather.

Distribution
Central and southern Africa
Size
30–55 mm (1¼–2¼ in) long

Pied fulgorid plant bug

Amantia imperatoria

This large fulgorid bug from Central America has bands and spots on the fore wings and the hind wings are red and brown. Fulgorid bugs are often attended by ants, which feed on the sticky honeydew excretions from their abdomens. These bugs are strong jumpers.

Distribution
Central America

Size
22–25 mm
(approximately 1 in) long

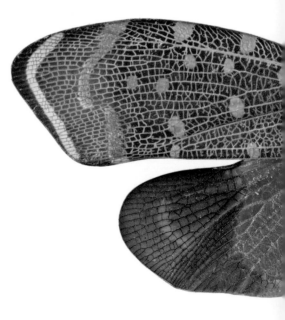

Flatid planthopper

Flatida rosea

As with other planthoppers, flatid bugs have sucking mouthparts and feed on the phloem sap of plants and trees. Most species cluster together on twigs in tropical forests. Adults of *Flatida rosea* have brightly coloured fore wings, unlike the membranous hind wings, which are used for flight. Adults of many flatids hold the wings above them like a tent, with the brightly coloured, hardened fore wings known as the 'tegmina'.

Distribution
Madagascar
Size
15 mm (¾ in) long

Claudina butterfly

Agrias claudina

This rainforest butterfly has vivid crimson patches on the upperwings, but the underwings are even more spectacular. The Claudina butterfly entranced British explorer and scientist Henry Walter Bates when he first encountered it in the 1850s in the Brazilian Amazon. It does, however, have some unsavoury feeding habits, as the butterflies often like to suck up nutrients from rotting flesh and fruit.

Distribution
Tropical South America

Size
80 mm (3¼ in) wingspan

Elephant hawkmoth

Deilephila elpenor

One of the more spectacular moths to be found commonly in gardens, the elephant hawkmoth combines lurid pink with green and black. The 'elephant' in its name derives from the caterpillar (not from the pink elephants in Dumbo!), which looks a little like an elephant trunk, although a lot more like a small snake. The caterpillars feed on willowherbs, plants fond of disturbed habitats, including around houses, so the adult moths are frequently seen at lights in suburban gardens.

Distribution
Europe and temperate Asia

Size
60–70 mm (2½–2¾ in) wingspan

Roseate emperor moth

Eochroa trimenii

For an emperor moth, this is a relatively small species, but it makes up for its size with its amazing pink wings and the male's outlandish feathery antennae. It lives in semi-desert habitat, the succulent karoo, of South Africa and is one of numerous habitat specialists being helped by conservation initiatives in Namaqualand, a region of western South Africa and Namibia with incredibly special and restricted vegetation types.

Distribution
South Africa

Size
65–75 mm (2½–3 in) wingspan

Picture-wing moth

Arniocera amoena

The picture-wing moths (Thyrididae) are found in tropical areas throughout the world. *Arniocera* species closely resemble the burnet moths of cooler northern areas, with gaudy red spots on a black background advertising the fact that these do not make a good dinner. This species was originally described as a burnet, an easy mistake to make.

Distribution
Southern Africa

Size
40 mm (1½ in) wingspan

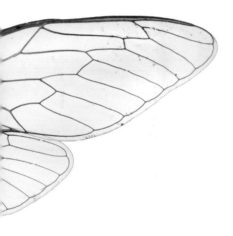

Caledonian cicada

Mouia variabilis

Cicadas are well known for the high-pitched droning calls of males. Cicadas are true bugs and feed on tree sap. The larvae develop below ground and feed from tree roots. Larval development can take several years to complete, dependent on the species, before the active larva crawls from the ground and pupates, usually on vegetation.

Distribution
New Caledonia

Size
25 mm (1 in) long

Atlas moth

Attacus atlas

Named after one of the Titans of Greek mythology, this is one of the largest of all moths, and females are larger than males, which is frequently the case in Lepidoptera. Not all names for this species reference its size, though. If you look at the extensions to the front sides of the fore wings, you'll see why it is known as the 'snake's head moth' in Hong Kong.

Distribution
South Asia

Size
250–300 mm (10–12 in) wingspan

Giant grasshopper

Tropidacris cristata

This tropical Central and South American species is the largest known winged grasshopper by length and wingspan, reaching up to 120 mm (4¾ in) and 230 mm (9 in), respectively. The fore wings resemble leaves and the hind wings are bright orange. The giant grasshopper is not often encountered in large numbers, but can be found in more open, drier situations in humid forests.

Distribution
Central and South America
Size
105–110 mm (approximately 4¼ in) long

Malagasy rocket moth

Epicausis smithii

The Malagasy rocket moth is the subject of a beautiful postage stamp issued in the Malagasy Republic. There are two species of rocket moth (genus *Epicausis*), found only in Madagascar, home to so many strange and unique animals. It is certainly a striking species within the noctuid (owlet) moths and known for the large crimson tuft at the end of the abdomen.

Distribution
Madagascar

Size
55–62 mm (2¼–2½ in) wingspan

Caliper beetle

Fruhstorferia nigromuliebris

The strange curved horns of this bright orange beetle are in fact their lengthened mandibles (upper jaws) and only appear in males. In contrast, the female is entirely black and has 'normal' mandibles. It is not known why colour varies so dramatically between the sexes but it is not unusual for caliper beetles. Little information is available about their ecology but deforestation of Asian rainforests for crops and timber will certainly be having a negative effect on the population of these beetles.

Distribution
Borneo

Size
25–44 mm (1–1¾ in) long

Jade headed buffalo beetle

Eudicella smithii

This species is popular among beetle breeders because of its rich variety of colour forms. The forebody can be reddish, green or blue and the elytra, hardened fore wings, vary from ochre to yellowish, with small or large black spots on the shoulders and on the rear. Only males possess the prominent, forked horn on the head, useful for pushing, shoving and prying off weaker rivals.

Distribution
Central and southern Africa

Size
25–40 mm (1–1½ in) long

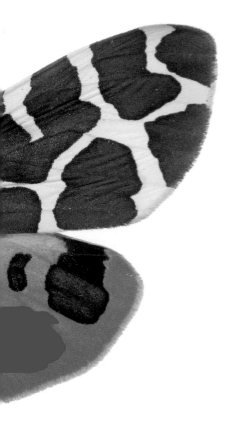

Garden tiger

Arctia caja

The strikingly coloured adults have warning spots on their hind wings and generally look unpalatable, which they are. Garden tigers contain compounds that are toxic to would-be predators. Often called 'woolly bears', the caterpillars are much less familiar than they used to be, at least in Europe. This is because the moth has suffered a steep decline in recent years, which may be linked to milder winters.

Distribution
Europe and North America
Size
45–65 mm (1¾–2½ in) wingspan

Orchid cuckoo bee

Exaerete trochanterica

These jewels of forests from Brazil to Panama are cuckoos in the nests of other orchid bees. They usurp the nests of *Eulaema* and *Eufriesea* bees, and are toughened to withstand their stings, which gives rise to the metallic cuticle. As with other orchid bees, males collect fragrant oils, particularly from orchids but also from other sources, even rotting wood, although we don't really know why.

Distribution
South and Central America

Size
23–26 mm
(approximately 1 in) body length

Hermann's tortoise beetle

Stolas hermanni

Alien in appearance, this round, smooth and flat tortoise beetle looks defenceless. However, if you were to try and remove it from a green leaf you would probably underestimate the strength of the grip! Tortoise beetles are known to secrete an oily substance from their feet that forms a film of liquid between the beetle and the surface it is standing on. This means it can withstand pulling forces several hundred times its own body weight.

Distribution
Amazon
Size
15–30 mm (¾–1¼ in) long

Oberthuer's fruit chafer

Mecynorhina oberthuri

This is a very powerful insect with a unique orange, white-and-black colour and a delicate covering of silky short hairs. Male fore legs are armed with large apical hooks and strong lateral spikes, which can easily slice through human skin, leaving a deep cut. They feed on tree sap and rotten fruits, like banana, often digging their bodies deep into the sugary mash. Females are smaller than males and hornless.

Distribution
Kenya and Tanzania
Size
44–80 mm (1¾–3¼ in) long

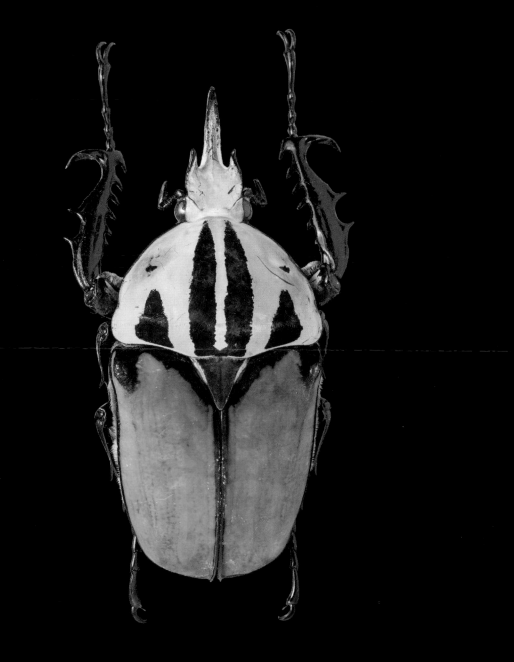

Chalcosiine burnet moth

Campylotes histrionicus

Campylotes histrionicus is a
conspicuous, day-flying moth that
can be found in late summer in
the foothills of the Himalayas and
other mountain ranges in southern
Asia. Like other burnet moths,
the chalcosiines are distasteful
to would-be predators, and they
employ active chemical defences
from glands. Many species are
involved in mimicry rings, where
various unrelated moths and
butterflies have evolved the same
colour pattern, so a species benefits
from appearing distasteful like its
mimic.

Distribution
Southeast Asia

Size
80 mm (3¼ in) wingspan

Jewel longhorn beetle

Sternotomis bohemani

This is a symmetrical gem of African rainforests and its impressive antennae contain microscopic sensors that can sense a freshly fallen tree from hundreds of metres away. As soon as an appropriate trunk is located, males take residence and wait for a mate. They are fiercely protective and guard their chosen partner from any potential rivals. After mating females lay elongated white eggs in the dead wood.

Distribution
Central and southern Africa
Size
20–25 mm (approximately 1 in) long

Leaf beetle

Platyphora princeps

Leaf beetles have many survival strategies. One is to emit a nasty smell, which is thought to keep enemies at a distance. Another tactic is to consume poisonous plants, which make the beetles taste bad. In addition, the strong contrasting colours of this beetle are a warning signal to potential predators – don't touch me! Some harmless leaf beetles know this trick and mimic poisonous relatives – the disguise can fool birds, but not spiders and other ravenous invertebrates.

Distribution
Mexico to Bolivia and Brazil
Size
15 mm (¾ in) long

Longhorn beetle

Trachyderes succinctus

The antennae of the longhorn beetles (often wrongly attributed as horns) are primarily used for sensing and are usually longer than the insect. Most longhorn beetles possess strong curved mandibles to bite into dead wood, which is where the females lay their eggs. The South American Titan beetle (*Titanus giganteus*) is often considered the largest insect (though not the heaviest, and not the longest including legs), with a maximum known body length of about 170 mm (6¾ in), also belongs in the same family, the Cerambycidae.

Distribution
Central and South America

Size
20–25 mm (approximately 1 in) long

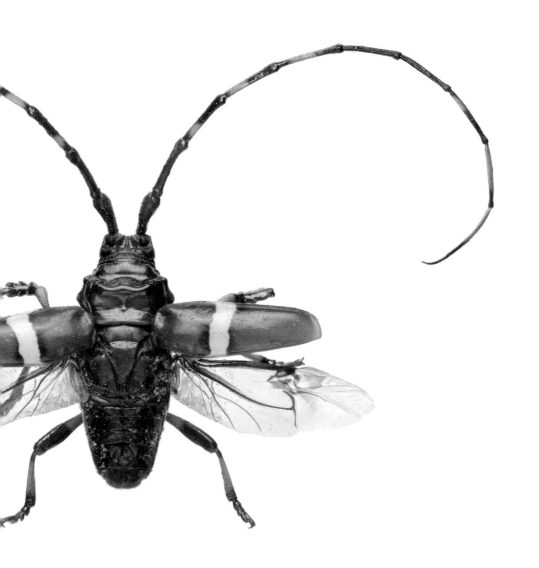

Common mapwing

Cyrestis maenalis rothschildi

Mapwing butterflies are so named because their wings seem to have lines of latitude and longitude drawn across them. As befitting a butterfly that seems to be mapping the globe, the common mapwing has a wide range in Southeast Asia, with many subspecies described. Far away in Europe, another nymphalid butterfly, *Araschnia levana*, has also been called the 'map'.

Distribution
The Philippines
Size
55–60 mm (2¼–2½ in) wingspan

Stalk-eyed fly

Achias rothschildi

An incredible looking fly with eyes literally at the end of stalks. Only the males of the species have long protrusions and they use them to show off to females who will judge them on length – a perfect indicator of strength and ability to defend a territory. Amazingly, stalk-eyed flies only develop the stalks when an adult emerges from the pupa – it swallows air through its mouth and pumps it through its eye stalks to inflate them!

Distribution
Papua New Guinea

Size
20–50 mm (1–2 in) distance between each eye

Giant cave cockroach

Blaberus giganteus

This is considered one of the largest cockroach species in the world. It is lightly built, with a flattened body, allowing it to hide in cracks from predators. At rest the wings are folded over the body, giving the cockroach a somewhat beetle-like appearance. It occurs in rainforests, and preferred habitats include areas of high moisture and low light, such as caves, tree hollows and cracks in rocks.

Distribution
Central America and northern South America

Size
65 mm (2½ in) long

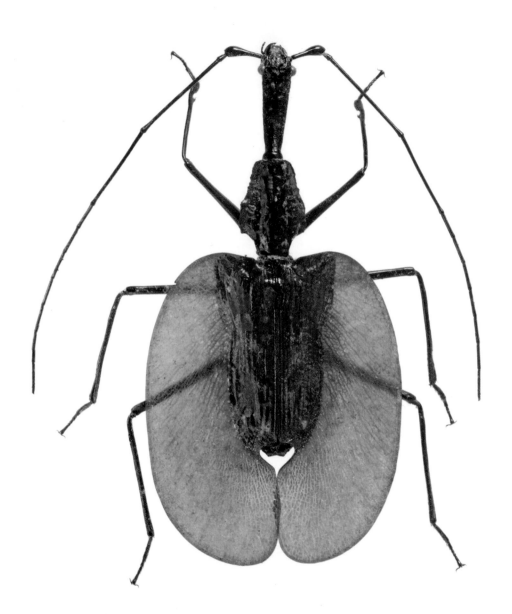

Violin beetle

Mormolyce phyllodes

This is perhaps the most unusual beetle in the hugely diverse family of ground beetles (Carabidae), of which six species, including this, occur in the Indo-Malayan Archipelago. The shape of its body has been compared to a guitar or violin and if viewed side on is completely flat! It is perfectly suited for its life under the loose bark of dead trees or in soil cracks. If disturbed, it will emit a spray of liquid from the tip of its abdomen. The fluid has a strong scent, resembling a mixture of nitric acid and ammonia, and causes a burning sensation if sprayed into eyes.

Distribution
Indo-Malaya

Size
60–100 mm (2½–4 in) long

Fairy fly (mymarid wasp)

Neomymar gusar

Fairy flies are actually parasitoid wasps and this species has huge, distinctive, black-tipped antennae to detect tiny traces of chemicals that their hosts emit. Mymarids parasitize the eggs of insects. As only a tiny wasp can develop in a tiny host, the undergrowth can be full of thousands of species of tiny parasitoid wasps. The micro-Hymenoptera specialist Alexandre Arsène Girault described them as 'gem-like inhabitants of the woodlands by most never seen nor dreamt of'.

Distribution
Central America

Size
3 mm (0.12 in) body length

Fulgorid bug

Penthicodes farinosa

This fulgorid bug occurs in the rainforests of Southeast Asia. The fore wings resemble tree bark or lichens and help the species to remain undetected on the trees on which they rest and feed. The species name *farinosa* refers to the wings looking as if they are dusted in flour.

Distribution
Malay Peninsula and Borneo
Size
15–20 mm (¾–1 in) long

Christmas beetle

Anoplognathus sp.

This beetle gets its common name because it is most active around Christmas, which is summertime in the Southern Hemisphere. Its life cycle is complex and takes several months. The larvae are soil-dwelling, the adults leaf-feeding. The thickened, robust legs and large claws of this specimen are thought to be used for holding onto thin branches.

Distribution
Australia
Size
20–30 mm (1–1¼ in) long

Excelsior eighty-eight

Callicore excelsior

Also known as the superb numberwing, *Callicore excelsior* is one of several numberwing species that seem to have numbers 'written on' by evolution. The upperwings contain red and blue, contributing to quite a rainbow of colours above and below. This is a butterfly with a huge appetite for the sweat of humans and other mammals, which it feeds on to obtain crucial salts.

Distribution
Tropical South America
Size
55–60 mm (2¼–2½ in) wingspan

Delias butterfly

Delias joiceyi

The underside of *Delias joiceyi* is
more extravagantly marked than
the upperside. Delias butterflies
are diverse in the Indo-Australasian
region. *Delias joiceyi* was named in
the early 20th century by George
Talbot after James John Joicey, one
of the most prolific collectors of
Lepidoptera from around the world.
Joicey's collection was one of the
major acquisitions that made the
Natural History Museum's collections
one of the largest in the world.

Distribution
Maluku (Spicy) Islands, Indonesia
Size
70 mm (2¾ in) wingspan

Bumblebee robber fly

Laphria flava

Robber flies or assassin flies are predators of other insects and use their stout beak (proboscis) to puncture the exoskeleton of other insects and suck out the contents. Many species mimic bees and use this disguise to actively hunt insects and deter would-be predators. This species resembles a bumblebee and mainly feeds on beetles.

Distribution
Europe, Asia, and North America
Size
12–25 mm (½–1 in)

Lime hawkmoth

Mimas tiliae

The lime hawkmoth is not named for its colour but for the fact that, in Britain, it almost always eats the leaves of lime trees. However, in other parts of Europe it happily eats those of birches too. One theory is that it takes longer for the larvae to develop when eating the relatively less nutritious birch leaves, so there isn't time in the short British summer to complete its life cycle.

Distribution
Europe

Size
55–70 mm (2¼–2¾ in) wingspan

Longhorn beetle

Cerosterna pollinosa

This beetle is a member of the megadiverse family of longhorn beetles (Cerambycidae) of which there are more than 26,000 species known worldwide. This bright yellow to pale orange, bulky bodied wood borer has long antennae and large sharp spikes either side of the thoracic shield (pronotum) used for defence. When disturbed it produces a chirping or creaking noise (called stridulation) by rubbing two parts of the body together. Male antennae are longer than females' antennae.

Distribution
Southeast Asia and Indo-Malaya
Size
45–65 mm (1¾–2½ in) long

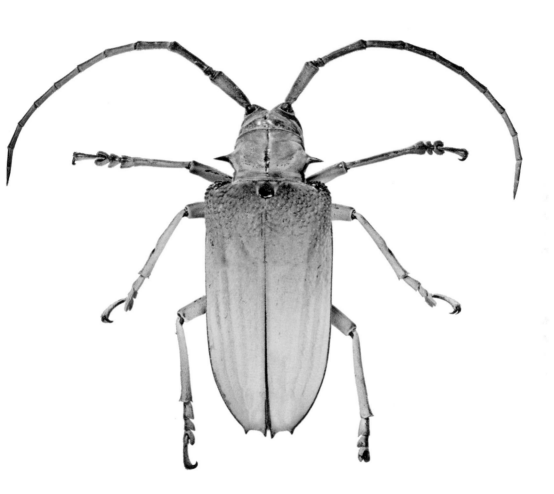

Chalcidid wasp

Conura dimidiata

Little *Conura* wasps are parasitoids of butterfly and moth pupae, or fly puparia in some species – the eggs are laid in the pupa and the wasp larvae eat the host alive. It is harmless to other creatures though, the sting wouldn't even penetrate skin. However, this chalcidid wasp mimics much more aggressive paper wasps, which certainly can sting.

Distribution
Colombia

Size
Body length 8–9 mm (¼–½ in) long

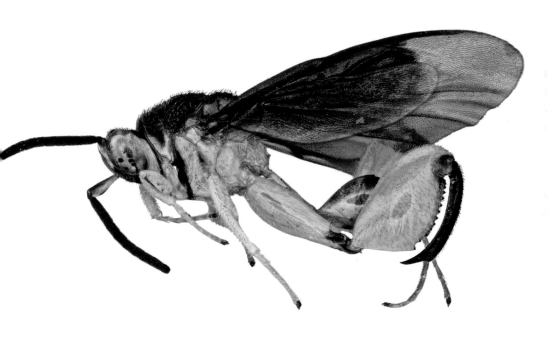

Tiger moth

Callindra principalis

One of many brightly coloured tiger moths, with flashy colours to warn of the adults' distastefulness and hairy caterpillars that most birds avoid. *Callindra principalis* is found in mountain ranges and was originally described in 1844 by an Austrian entomologist from specimens that returned to Europe from an expedition, at a time when many Asian species were first seen by Europeans.

Distribution
Asia, Pamir and Himalayan mountain ranges
Size
80 mm (3¼ in) wingspan

Green milkweed grasshopper

Phymateus viridipes

This large African grasshopper secretes a noxious fluid from the thorax when alarmed. The fluid is derived from the poisonous milkweed plants upon which they feed as nymphs and adults. The coloured hind wings, which are normally hidden when the grasshopper is at rest, can also be flashed to deter potential predators.

Distribution
Southern Africa

Size
70 mm (2¾ in) long

Wallace's golden birdwing

Ornithoptera croesus

Alfred Russel Wallace captured this spectacular butterfly in Maluku, Indonesia, and it is fair to say that he was rather excited by this discovery. Wallace documented how the thrill caused the blood to rush to his head, he felt faint and developed a cracking headache. Later, Wallace described this species as new, but he is better remembered as the co-originator of the idea of evolution through natural selection.

Distribution
Moluccan (Spicy) Islands of Indonesia
Size
130–190 mm (5¼–7½ in) wingspan

Jewel scarab beetle

Chrysina sp.

If glimpsed in the sunlight, this jewel scarab beetle's brilliant metallic gold colouring shines like a Christmas decoration. The high reflectiveness of the exoskeleton in this group of ornate beetles manipulates a property of light called polarization, and the way the light is reflected makes its hard case look like pure gold. The larvae develop in rotten wood, while the adults commonly feed on foliage and are active in daytime.

Distribution
Southern USA to northern South America
Size
18–30 mm (¾–1¼ in) long

Ichneumonid wasp

Xanthopimpla summervillei

Throughout tropical regions, but especially Asia, hundreds of species of bright yellow and black *Xanthopimpla* dart about forests and undergrowth, seeking moth pupae to parasitize. *Xanthopimpla* have peculiar big bristles on their claws, which might act as 'poison fangs' to deter predators. The patterns of black dots and stripes can be distinctive for different species, although identifying these vivid little wasps is surprisingly difficult, as there are so many species.

Distribution
Southeast Asia and Australia
Size
10–11 mm (approximately ½ in) long

Aumakua

Aumakua omaomao

This distinctive pink moth is the only species of its genus, and it is restricted to the remote Hawaiian islands. Caterpillars feed on flowering plants called lobelioids (bellflowers), which produce indigestible latex. The caterpillar cuts the plant then returns to eat when the latex has leaked out.

Distribution
Hawaii

Size
35–40 mm (approximately 1½ in) wingspan

Yellow umbrella stick insect

Eurynecroscia nigrofasciata

A predominantly ground-living stick insect that occurs in tropical forests. Males are extremely rare and are considerably smaller than the females (55–59 mm; approximately 2¼ in). Most stick insects are capable of asexual reproduction, without fertilization by the male, and females lay up to 100 eggs individually. They make popular pets in parts of Southeast Asia.

Distribution
Borneo, Peninsular Malaysia and Sumatra
Size
77–97 mm (3–3¾ in) long

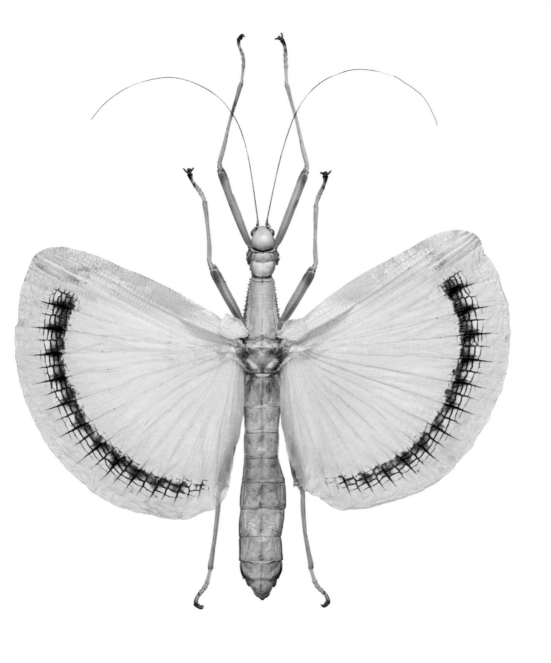

Pale horse fly

Cryptotylus unicolor

Adult horse flies feed on blood and some species are known to spread diseases to livestock. The larvae have very different habits, however, and normally develop in mud around ponds and streams or in the rot holes of forest and savanna trees. The adult female fly normally requires a blood meal before the eggs will mature. This species of medium-sized horse fly occurs in the tropical forests of South America.

Distribution
Tropical South America

Size
18–20 mm (¾–1 in) long

Coffee bee hawkmoth

Cephonodes hylas virescens

This is one of several hawkmoths that mimic bees, and it has a huge range across warmer parts of the Old World. The similarity to bees is emphasized by the lack of scales on the wings. These moths hover in front of flowers, feeding on nectar with their long proboscis, and probably not contributing much to pollination.

Distribution
Africa

Size
55–62 mm (2¼–2½ in) wingspan

Luna moth

Actias luna

Named after Luna, the Roman moon goddess, by Linnaeus, this is one of the largest moths in North America. It is a giant silk moth with long tails that confuse bat sonar. As with other silk moths, adult luna moths don't feed, so they spend a relatively brief amount of time as an adult, with males sniffing out females using their large feathery antennae.

Distribution
North America

Size
110–180 mm (4¼–7 in) wingspan

Green paper wasp

Belonogaster prasina

Belonogaster is a genus of many species of paper wasps, but the only green species are found in Madagascar. Why that should be, we don't know, but they are certainly well camouflaged when hanging from their nests in the vegetation. Colonies are relatively small and founded by a single queen. The stings of these paper wasps are fierce, necessary to defend their nest, which is an open layer of paper cells, lacking a protective outer envelope.

Distribution
Madagascar
Size
30 mm (1¼ in) long

Emperor dung beetle

Sulcophanaeus imperator

These dung beetles tend to form a close association between the male and female (a pair bond). Together they excavate and build feeding galleries, where they store the dung they have collected to feed on, even before the female is sexually mature and ready to reproduce. Surprisingly colourful, this dung beetle inhabits dry forests, scrublands and pastures and is one of the most common insects in cattle-raising areas of southern South America, with a crucial role in agricultural ecosystems.

Distribution
South America

Size
13–25 mm (½–1 in) long

Rothschild's birdwing

Ornithoptera rothschildi

Like all the birdwings, Rothschild's birdwing is a spectacular butterfly, with females much larger but less colourful than males. The underside (pictured) is even more spectacular than the upperside. They are found only in meadows fairly high in the Arfak Mountains of West Papua. This birdwing is named after Lord Walter Rothschild, who financed the expeditions that discovered this elusive species. Rothschild's huge collection of butterflies now resides in the Natural History Museum, London.

Distribution
West New Guinea
Size
130–160 mm (5¼–6¼ in) wingspan

Verdant hawk

Euchloron megaera

A very widespread hawkmoth in Africa, *Euchloron megaera* undertakes migrations within Africa and has established distinctive populations on some Indian Ocean islands. When Linnaeus described this distinctive green hawkmoth in 1758, he thought the specimen was from India. Accurate labels are very important in taxonomy.

Distribution
Africa south of the Sahara, Yemen and Indian Ocean

Size
95–120 mm (3¾–4¾ in) wingspan

Flower chafer

Protaetia aeruginosa

This European beetle is endangered due to habitat changes and the disappearance of old hollow trees in which it prefers to develop. In a completely natural ageing process, at some point in their life, deciduous trees develop internal hollows. Sadly, many people consider such hollow trees to be 'sick' and remove them. The habitat of hundreds of species disappears with one felled tree.

Distribution
Central and southern Europe
Size
up to 34 mm (1¼ in) long

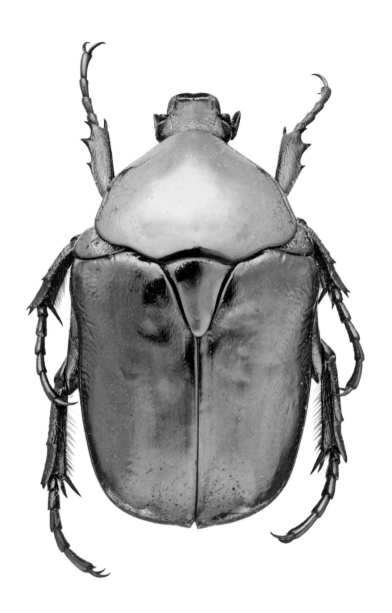

Kaiser-i-Hind

Teinopalpus imperialis

This unmistakable and unusual swallowtail butterfly is found at high elevations in mountain ranges, particularly the Himalayas. Its common name translates as the 'emperor of India'. It seems to fly only in the morning, when it tends to stay high up in the trees and fly fast. Despite the elusive nature of this butterfly, there has been much research conducted on its brilliant green, refractive wing scales.

Distribution
Southeast Asia

Size
90–127 mm (3½–5 in) wingspan

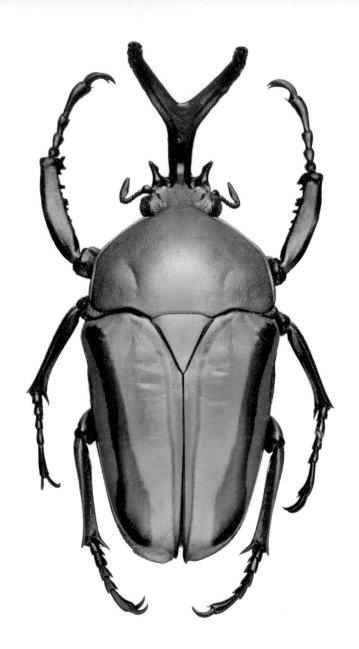

Striped love beetle

Eudicella gralli

There are over 4,000 species of flower chafers. Depicted here is one of the colour morphs of the highly variable, striped love beetle. The patterning on these insects can vary from almost entirely emerald green to reddish-orange, with the hardened wings longitudinally striped in green, orange or black. Because of the variability in colour, it is obvious that visual recognition of a partner or rival is difficult, so chemical signals are widely used.

Distribution
Equatorial Africa
Size
22–45 mm (1–1¾ in) long

Flower chafer

Argyripa gloriosa

The Latin epithet of the species name of this beetle refers to its spectacular shining green and yellow colouration, which is intermixed with five pairs of irregularly shaped black spots on the dorsal surface. Warmer tones are aligned along the middle, while darker tones are more saturated on either side. Rarest in its genus, this chafer was discovered and classified scientifically only in 1978.

Distribution
Colombia

Size
19–21 mm (approximately 1 in) long

Green oakblue

Arhopala eumolphus

This little hairstreak butterfly has striking differences between the sexes. Males are metallic green above whereas females are black and metallic blue. Many lycaenid butterflies, including the hairstreaks, have little tails on the hind wings. These are coupled with small eye spots on the underside of the wing, fooling would-be predators into attacking the end of the wing, where damage matters a lot less than attacks to the head end of the butterfly.

Distribution
Indo-Malaya

Size
40–44 mm (1½–1¾ in) wingspan

Tortoise beetle

Cyclosoma palliata

Round and flat, this beetle resembles a saucer placed upside down. Adults and their larvae are plant-eaters. Tortoise beetles' larvae are enigmatic in their appearance of spiky, trilobite-like body with an intricate long, feathery tail. Scared larvae flick up their tail and cover their body with this feathery mass. The 'tail' is, in fact, not a part of the living organism, but the animal's dried faeces used to camouflage the body.

Distribution
French Guiana

Size
10–13 mm (approximately ½ in) long

Gray's leaf insect

Phyllium bioculatum

Leaf insects are stick insects that have evolved extremely flattened, irregularly shaped bodies, wings and legs. They are called leaf insects because the large, leathery fore wings of the females have veins that closely resemble the veins of leaves, giving them superb cryptic camouflage. Adult male leaf insects have transparent wings and in this species conspicuous spots on their abdomen, hence their scientific name meaning 'two-spotted'.

Distribution
Southeast Asia and Indo-Malaya

Size
50–100 mm (2–4 in) long

Emerald swallowtail

Papilio palinurus

A common species of forest clearings, the emerald swallowtail butterfly's brilliant green has made it a popular inclusion in butterfly houses. Like many butterflies described in the early days of entomology, the species name has a classical allusion, referring to the pilot of Virgil's ship in the Aeneid. The vivid green is produced by wing scales like artists' palettes, refracting blue and yellow wavelengths of light, mixing to form green.

Distribution
Southeast Asia

Size
75–100 mm (3–4 in) wingspan

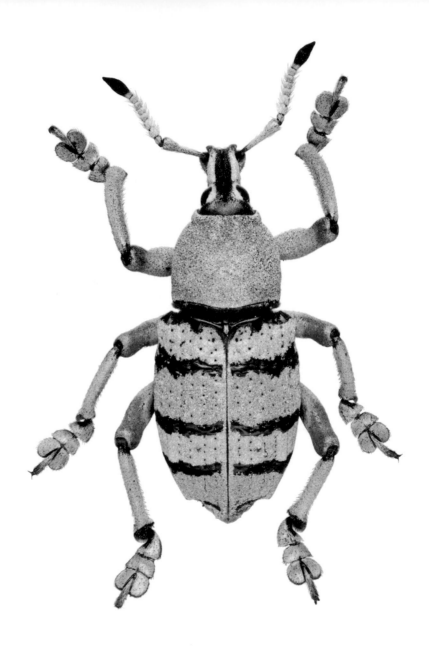

Papuan green weevil

Eupholus schoenherrii

Beetles of the genus *Eupholus* are rightly considered the most beautiful weevils. Although brightly coloured, their colouration is in fact cryptic, combining the blue tropical sky, green lush vegetation and the darkness of tropical rainforests. This particular species is not uncommon all through northern New Guinea and inhabits both primary forests and local gardens.

Distribution
New Guinea
Size
21–34 mm (¾–1¼ in) long

Southern green stink bug

Nezara viridula

Stink bugs are true bugs that emit a nasty smell when disturbed or agitated. They feed on plant sap as nymphs and adults and some are crop pests in various parts of the world. The southern green stink bug has spread throughout the world and is now almost cosmopolitan. Adults are often attracted to the lights of houses.

Distribution
Virtually cosmopolitan (originally African)
Size
10–15 mm (½–¾ in) long

Oryba sphinx moth

Oryba kadeni

The extremely large eyes, short antennae and bright green colouration are distinctive features of this widespread Neotropical hawkmoth. The huge eyes are able to take advantage of any available light, and the brain of this moth also processes what it sees more slowly, which helps the moth to see more detail in the dark.

Distribution
Central and northern South America

Size
102–116 mm (4–4½ in) wingspan

Madagascan sunset moth

Chrysiridia rhipheus

Sunset moths are often mistaken for butterflies, and *Chrysiridia rhipheus* was originally described as a *Papilio* (swallowtail) species in 1773. The flashy colours of this ornate moth indicate that it is poisonous, and the caterpillars do indeed feed on toxic *Omphalea* plants. This is a migratory moth, moving from east to west across Madagascar to find new food sources. Its food-plant becomes more toxic the more that caterpillars feed on it, which sends the moths on a restless quest to find more palatable food where there are currently no caterpillars.

Distribution
Madagascar

Size
70–110 mm (2¾–4¼ in) wingspan

Orchid bee

Euglossa intersecta

Like many other bees, orchid bees collect pollen and nectar from plants – not just orchids, although males are especially attracted to the scent of orchids. Males also collect scents in special pouches on their hind legs that might help attract a mate. They are powerful fliers and travel amazing distances, sometimes up to 48 km (30 miles) a day, in their quest for the perfect mixture of smells.

Distribution
Tropical South America
Size
20 mm (1 in) long

Spotted flower chafer

Stephanorrhina guttata

This jungle beauty is brilliant metallic green and red, with white spots on the hardened fore wings. The adults feed on nectar and are often observed in large numbers on flowering bushes; the characteristic buzzing of their wings can be heard from a distance. Males are somewhat smaller than females. The C-shaped, fat, white larva lives in and feeds on decomposing wood. A popular pet, this beetle successfully breeds in captivity.

Distribution
Equatorial Africa

Size
20–25 mm (approximately 1 in) long

Magnificent weevil

Eurhinus magnificus

Imagine sitting on the same sofa for your whole life? Like many plant-eaters, the magnificent weevil is described as a narrow specialist, as its whole life cycle takes place on just one species of tropical vine, *Cissus verticillata*. It takes as long as 83 days to develop from an egg to adult beetle. This originally Central American species was first discovered in Florida in 2002.

Distribution
Central America to Mexico, introduced into southern USA

Size
5.5–5.8 mm (approximately ¼ in) long

Rajah Brooke's birdwing

Trogonoptera brookiana

The spectacular Rajah Brooke's birdwing is declining due to habitat destruction and over-collecting for trade, despite being protected. Males love to visit mineral-rich puddles, which makes them easy to find and count, to monitor the strength of populations. Females are more difficult to find, flying higher up and at higher elevations. This is one of many spectacular species first collected and described by Alfred Russel Wallace in his lengthy explorations of Southeast Asia.

Distribution
Malay Peninsula and Sumatra

Size
150–170 mm (6–6¾ in) wingspan

Tortoise beetle

Omocerus azureicornis

A green plant-eater using a green leaf for camouflage? Why not? This is a very common strategy in the world of insects. However, this emerald creature has another tactic up its sleeve! Owing to the differences in structure between insect and bird eyes, this shiny emerald creature can evade detection by insectivorous birds but is clearly visible to rivals and partners.

Distribution
Central and South America
Size
12–15 mm (½–¾ in) long

Golden-green weevil

Baris cuprirostris

One of the puzzles of life is why many insects are common and annoying, and others are naturally rare. This little marvel feeds on a wide range of herbs, including mustard, cabbage, woad and other crucifers. However, because of its rarity, it is never considered a pest. Adult beetles emerge from pupal chambers at the end of the summer and spend the winter in the soil near the host plant.

Distribution
Europe except in the east and north, northern Africa

Size
2.5–3.5 mm (0.10–0.14 in) long

Spotted tachinid fly

Formosia moneta

Some of the most impressive tachinid flies occur in New Guinea and the Pacific Ocean islands. This species has a bright metallic thorax and iridescent spots on the abdomen that reflect light. Although these flies superficially resemble 'bluebottle' flies they are in an entirely different family, with different larval habits and biology.

Distribution
Papua New Guinea

Size
15 mm (¾ in) long

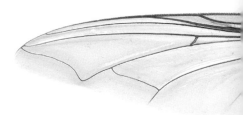

Cuckoo wasp

Chrysis ruddii

The cuckoo wasps live up to their name by laying their eggs in the nests of bees and wasps, with *Chrysis ruddii* specializing in the clay nests of potter wasps. The young cuckoo wasp eats the rightful occupant and then the food store. When attacked by the bees or wasps they are trying to usurp, cuckoo wasps can roll up into a heavily armoured, jewel-like ball.

Distribution
Throughout Europe and western Asia
Size
7–10 mm (¼–½ in) long

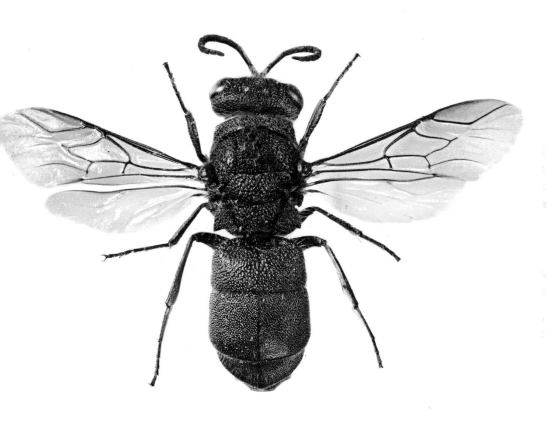

Chalcosiine moth

Eterusia aedea

Eterusia aedea is a burnet moth in the subfamily Chalcosiinae, a group of beautifully patterned moths with complicated chemical defences and communications. But not everybody is an admirer of this green, blue and white moth. Its spiky, hairy larvae, the 'red slug caterpillars', can be major pests of tea, munching through leaves in tea plantations in China, India, Sri Lanka and other countries.

Distribution
Southeast Asia

Size
50–70 mm (2–2¾ in) wingspan

Eucharitid wasp

Chalcura cameroni

Eucharitid wasps parasitize ant larvae, with the wasp larva hitching a ride into the nest on a worker ant by grabbing on with its mandibles as the ant walks past. Adult wasps are tough, with projections on the body to allow ants to grab them and remove them from the nest, without the wasps coming to harm. The hardened insect cuticle often tends to be metallic too, as in this case.

Distribution
Southeast Asia and Australia
Size
3.6 mm (0.14 in) body length

Emerald cockroach wasp

Ampulex compressa

The beautiful *Ampulex* wasp has a rather gruesome biology, but one that many of us can appreciate. It hunts cockroaches, using potent and unusual venom to subdue its prey. The venom is injected precisely into the brain of the cockroach, with the effect that the cockroach loses its willpower. Instead of escaping it is willingly walked by the wasp to the wasp's nest burrow, where it becomes food for the *Ampulex* larva.

Distribution
Widespread in tropical Southeast Asia
Size
18–22 mm (¾–1 in) long

Queen Alexandra birdwing

Ornithoptera alexandrae

Restricted to pristine rainforests of Oro Province in Papua New Guinea, *Ornithoptera alexandrae* is the largest butterfly known, and one of the most threatened. Although it is relatively numerous in its small range, its forest habitat is threatened by logging and volcanic eruptions. This is also a highly coveted butterfly, fetching huge prices, resulting in trade being totally banned. The original (type) specimen of *Ornithoptera alexandrae* is peppered with small holes, from when it was blasted from the sky with a shotgun!

Distribution
Papua New Guinea

Size
160–250 mm (6¼–10 in) wingspan

Pteromalid wasp

Chalcedectus maculicornis

Chalcedectus maculicornis belongs to a group of small, often brightly metallic-coloured, parasitoid wasps that are capable of powerful jumps, many times their body length. Specimens in collections are often strongly bent, which is how they jump: the thorax clicks backwards and they launch themselves by releasing a muscle. The muscle contains resilin, a strongly elastic protein, which enables wings to beat, fleas to leap, and many other insect feats.

Distribution
Brazil
Size
6 mm (¼ in) body length

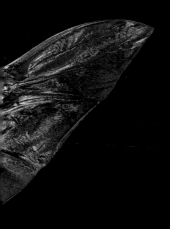

Goliath beetle

Goliathus goliatus

Goliath beetles and their close relatives the rose, fruit and flower chafers, are characterized by their distinctive flight. While most beetles that are able to fly will broadly open both pairs of wings, exposing their vulnerable soft abdomen, Goliath beetles release their flying wings from under closed, protective, hardened fore wings. Despite being bulky and heavy, these beetles are powerful fliers, commonly found around the canopies of blossoming trees.

Distribution
Equatorial Africa

Size
50–110 mm (2–4¼ in) long

Green dock beetle

Gastrophysa viridula

Until the end of the 19th century
the green dock beetle was limited to
mountainous areas and fed mainly
on alpine dock. However, extensive
use of fertilizers in European
agriculture has meant that dock
species have spread widely in
pastures and meadows, and now the
beetle is very common across central
and northern Europe. The black
larvae have glands that produce
a variety of irritating chemicals.

Distribution
Eurasia (not in the south), North America
Size
3.5–7 mm (0.14–¼ in) long

Frog-legged leaf beetle

Sagra buqueti

The males of the iridescent frog-legged beetle have monstrous hind legs. And, despite what its name suggests, it doesn't use them for jumping like a frog. Instead, the bizarre legs help it to cling onto stems and foliage while the herbivorous beetle eats. The grip is aided by tiny hair follicles that cover the surface of the leg. Female hind legs aren't as big.

Distribution
Southeast Asia
Size
23–50 mm (1–2 in) long

Oak gall ormyrus

Ormyrus nitidulus

Gall wasps induce plants to produce an excess growth of nutritious tissue, which their larvae then eat. This growth is called a gall, and it also offers some shelter from predators. However, many other insects have adapted to take advantage of the shelter and food provided by galls, including *Ormyrus nitidulus*. Large galls on oaks often harbour *Ormyrus*, which have previously paralyzed and laid an egg on the insects that formed the galls.

Distribution
Europe

Size
3–5 mm (approximately ¼ in) body length

Metallic tachinid fly

Rhachoepalpus metallicus

As its scientific name suggests, this fly has a striking metallic blue sheen and its abdomen is clothed in long stout, erect bristles. Metallic colouration is not very common in this family of flies, but several metallic species are known from different regions of the world. The species occurs in the High Andes of tropical South America, where the larvae probably develop as internal parasites of caterpillars or beetle larvae.

Distribution
Tropical South America

Size
12–15 mm (½–¾ in) long

Imperial arcas

Arcas imperialis

Males of *Arcas imperialis* are very faithful to particular resting places in the forest. They vigorously defend their territory and always return to their favoured perch. These perches also act as meeting places for males and females. Although some *Arcas* species seem to prefer intact forest, *Arcas imperialis* has adopted disturbed habitats, including around towns, which is good news for this beautiful butterfly.

Distribution
Central and South America
Size
33–37 mm (1¼–1½ in) wingspan

Papuan green weevil

Eupholus bennettii

In its natural environment, you only have a chance to observe this extraordinary weevil in the eastern part of the world's largest tropical island, New Guinea. It is locally common on certain species of vine. These weevils are caught by Papuans and used for decoration in necklaces and earrings at their magnificent tribal celebrations.

Distribution
New Guinea

Size
22–35 mm (1–1½ in) long

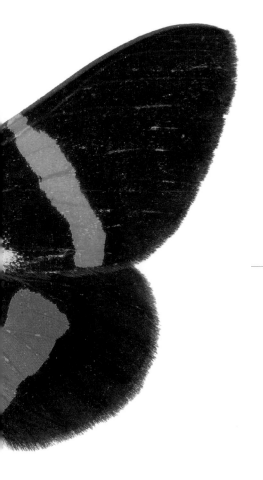

Geometrid moth

Milionia weiskei rubidifascia

Feeding on conifers in New Guinea, moths of the genus *Milionia* are among the most colourful geometrid moths. These are day-flying, conspicuous insects, often to be seen visiting flowers, rotten fruit or dung. This particular subspecies gets its name, *rubidifascia*, from the bold red stripe on its hind wing.

Distribution
New Guinea

Size
50–56 mm (2–2¼ in) wingspan

Blue carpenter bee

Xylocopa caerulea

Whilst the carpentry of carpenter bees is limited, these bees do chew impressively large holes in dead wood in which to make their nests. They are not truly social insects – each female raises her own offspring – although they often share entrances to their individual nest chambers. Many species of *Xylocopa* are impressively large and shiny, with iridescent wings, although *Xylocopa caerulea* is the only one with a blue fur coat.

Distribution
Widespread in Southeast Asia

Size
23–28 mm (1–1¼ in) body length

Procilla beauty

Panacea procilla

These stunning forest butterflies aren't to be found on flowers. Instead, they love to visit fruit to obtain their sugars. They can also be seen basking on the trunks of trees with their wings widespread. There is no need for a butterfly to hide away under camouflaged underwings when it is as toxic as *Panacea procilla*.

Distribution
Central and South America

Size
80–95 mm (3¼–3¾ in) wingspan

Brush jewel beetle

Julodis cirrosa

This metallic blue-green jewel beetle has a cylindrical body with a surface coarsely punctured and covered with tufts of long whitish, yellow or orange wax-coated hairs. It is not uncommon in dry and semi-dry warm regions in southern Africa. The larvae tunnel in stems and roots of various shrubs. The adult beetles are short-lived, active during the heat of the day and feed on water-rich foliage and flowers.

Distribution
Southern Africa
Size
25–27 mm (approximately 1 in) long

Chalcosiine day-flying moth

Erasmia pulchella

As with many other burnet moths, *Erasmia pulchella* has a natural chemical defence in the form of hydrogen cyanide. This means it doesn't need to waste energy flying fast or finding somewhere to hide. This is a conspicuous day-flying moth with a rather fluttery flight. It is found in forests, where the caterpillars feed on a flowering shrub, *Helicia cochinchinensis*, this is an unusual family of plants for this moth to feed on.

Distribution
Southern and eastern Asia

Size
70–80 mm (2¾–3¼ in) wingspan

Bates's tortoise beetle

Discomorpha batesi

Surprisingly, some tortoise beetles show maternal care of offspring, which is uncommon in the insect world. Females have been reported guarding their larvae and pupae. Sometimes both male and female take care of the larval group. Look carefully on this seed-like rainforest gem and discover the apparent outline of another, smaller tortoise beetle on the back of this specimen.

Distribution
Amazon
Size
9–12 mm (¼–½ in) long

Giacometti's wasp

Umanella giacometti

How do we concoct names for newly
discovered species? One answer is
to hold a competition, which is how
this wasp earned its name in 2010.
The suggestion that this elongate
wasp resembled the figures of the
Swiss sculptor, Alberto Giacometti,
won the day. Presumably the bright
metallic colours are warnings to
predators not to touch it whilst it
probes dead wood for beetle grubs.

Distribution
Ecuador and Peru

Size
71–79 mm (2¾–3¼ in) long

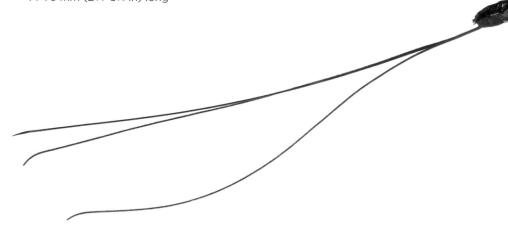

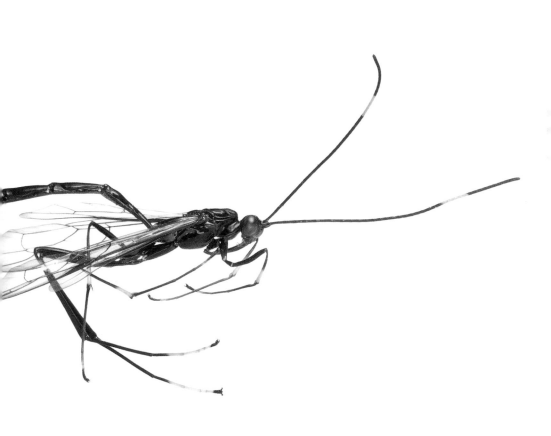

Blue morpho

Morpho cypris

Coloured in brilliant metallic blue and silvery-white, is considered by many to be the most beautiful butterfly in the world. The word Cypris (from Cyprian) is another name for Aphrodite, the Greek goddess of love and beauty. The undersides of the wings are light brown and whitish. The blue morpho is a butterfly that requires mature jungle canopy to flourish and is most frequently encountered along waterways.

Distribution
Central and South America

Size
130–200 mm (5¼–8 in) wingspan

Red speckled jewel beetle

Chrysochroa buqueti

Perhaps the most rapid fliers of all beetle species, these glorious insects also have large eyes that cover the sides of their head, so it is difficult for many birds to get close enough to catch them without being noticed. Unfortunately for jewel beetles, because of their extraordinary beauty they have become popular souvenirs for people and are caught and framed to sell to tourists in many Asian nations.

Distribution
Southeast Asia
Size
40–50 mm (1½–2 in) long

Orchid cuckoo bee

Aglae caerulea

Most orchid cuckoo bees belong to the genus *Exaerete*; however, *Aglae caerulea* is one species in its own genus. These bright blue bees have a tough cuticle that helps when being attacked by the rightful owners of the nests they are taking over. As with other orchid bees, males like fragrant oils and can easily be attracted by artificial baits of methyl cinnamate, used in artificial flavours and perfumes.

Distribution
South America
Size
23 mm (1 in) body length

Brazilian jewel beetle

Conognatha haemorrhoidalis

First described in 1790, this deep blue reflective and shiny jewel beetle is rare. Its hardened fore wings (scientifically called elytra) have little spines that run along the sides of the posterior half of the body, but the function of the spines remains unclear. If not able to fly away from a potential predator, the beetle plays dead and keeps all its legs close and tucked under its body as shown here. The shiny elytra of many jewel beetles are used in jewellery and decoration in certain countries.

Distribution
Brazil
Size
30 mm (1¼ in) long

Blue ground beetle

Carabus intricatus

The blue ground beetle is flightless
– it has no hind wings, which are
commonly used for flight in other
insects – however, what it lacks in
flight it makes up in speed and is a
tireless runner making it a ferocious
predator of slugs and earthworms.
The long, slender larva is armed with
large, sharp mandibles and searches
for its prey in soil and under logs.
Unfortunately, this long-lived beetle
is becoming increasingly rare and is
threatened with extinction in some
countries.

Distribution
Europe
Size
24–35 mm (1–1½ in) long

Claudina butterfly

Agrias claudina

The Claudina butterfly earns its appearance twice in this book, as the upperwings are as vividly colourful as the underwings are intricately patterned. The hind wing has yellow tufts, which are called the androconia. These are patches of special scales on many male Lepidoptera. Their purpose seems to be to diffuse pheromones, which are involved in courtship.

Distribution
Tropical South America
Size
80 mm (3¼ in) wingspan

Darkling beetle

Odontopezus sp.

This species from Africa has a purple sheen – a striking and uncommon feature among its generally dark relatives. Compared to most insects, many darkling beetles are long-living and adults can reach over two years of age, thanks to their slow metabolism. The adults of darkling beetles usually have a chemical defence mechanism and are therefore relatively protected against predators. Some species of this megadiverse family (Tenebrionidae) have been spread across the globe on products such as timber.

Distribution
Western and Central Africa
Size
40–50 mm (1½–2 in) long

Bird-locust

Ornithacris pictula magnifica

This species derives its common name from its bird-like flight when disturbed. The fore wings resemble leaves and help to camouflage the insect when at rest, but the hind wings are reddish-purple and probably help to deter predators when the locust takes to flight. Bird-locusts are important crop pests in Africa and have been recorded attacking oleander, palms, maize, millet and other crops.

Distribution
Sub-Saharan Africa

Size
50 mm (2 in) long

Andromeda satyr

Cithaerias andromeda

Cithaerias butterflies are elusive rainforest dwellers, typically found in the darkest undergrowth and flying in the evening. The transparent wings make them even more difficult to spot and add to their ghost-like nature; indeed, these butterflies are sometimes referred to as 'phantoms'. They emerge to feed on rotting fungi and fruits.

Distribution
Tropical South America

Size
35–48 mm (1½–2 in) wingspan

Egyptian scarab

Scarabaeus aegyptiorum

This dusk purple dung beetle uses its strong wide head as if it was a spade. It digs into fresh cattle dung, cuts off a piece and rolls it some distance from where it was deposited before burying it in underground chambers. Beetles then consume the dung or the females lay eggs on it. Growing larvae feed on the dung ball and pupate in the chamber. Adults spiked front legs are useful for gripping, rolling balls and digging them into the ground.

Distribution
North-eastern Africa
Size
22–35 mm (1–1½ in) long

Mottled horse fly

Euancala maculatissima

This beautifully marked African species of horse fly probably feeds on antelopes. Horse flies suck the blood of mammals, reptiles and amphibians, but very few have been reported as attacking birds. Some species, however, have extremely long mouthparts and are adapted to feed on the nectar of flowers. This species occurs in forests and dry savanna habitats.

Distribution
Africa
Size
12 mm (½ in) long

Speckled emperor

Gynanisa maja

The feathery antennae of male speckled emperors are for detecting pheromones, which the females emit in the middle of the night. The caterpillars feed mainly on *Acacia* and mopane. In turn, people eat the caterpillars of *Gynanisa maja* in large numbers, together with those of a few other species of Lepidoptera. This attractive moth even features on a postage stamp in Mali.

Distribution
South and East Africa

Size
105–113 mm (approximately 4¼ in) wingspan

Dead leaf mantis

Deroplatys truncata

Mantises are a diverse group of predatory insects with many species that are cryptically camouflaged to deceive their prey. Some are almost flawless mimics of flower petals, bark and lichens. This Asian species hunts on the forest floor and is beautifully adapted to resemble dead leaves. It uses its spiny raptorial front legs to grasp and impale insect prey.

Distribution
Indo-Malaya and New Guinea
Size
65 mm (2½ in) long

Lantern bug

Zanna nobilis

This lantern bug is mostly grey with black speckling. The head is developed into a long 'snout' with folds or protuberances on the surface. The fore wings help to camouflage the bugs against tree bark, and the hind wings are unmarked. These bugs occur in Southeast Asia, where they feed on tree sap. *Zanna* nymphs in Madagascar are considered a delicacy and are supposed to taste like bacon.

Distribution
Peninsular Malaysia and Borneo

Size
50–55 mm (2–2¼ in) long

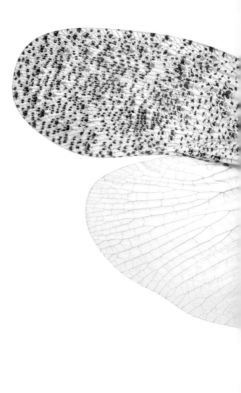

Bubble-winged darkling beetle

Metallonotus physopterus

There are over 20,000 species in the cosmopolitan megadiverse family of darkling beetles (Tenebrionidae). They appear in all types of terrestrial habitats from deserts to rainforests and inhabit altitudes from sea level to the montane snowline. This ornate flightless African rainforest dweller has hardened wings that are coarsely punctate and unusually pear-drop in shape. Mainly nocturnal, these insects can be found on rotten wood with tree fungi, where opposite genders communicate through chemical signals.

Distribution
Western and Central Africa
Size
Up to 30 mm (1¼ in) long

Three-horned rhinoceros beetle

Chalcosoma moellenkampi

The two horns on the pronotum and one on the head of the male three-horned rhinoceros beetle are marvellous. However, depending on environmental conditions, not all larvae have enough food, therefore many adult beetles are small with rudimentary horns. Large males are powerful and very aggressive towards their rivals and fierce fights occur regularly. Massive larvae of 100 g (3.5 oz) weight inhabit rotten wood, grow over a year, are important decomposers of dead wood and are known to bite when disturbed.

Distribution
Borneo

Size
Males up to 110 mm (4¼ in), females up to 60 mm (2½ in) long

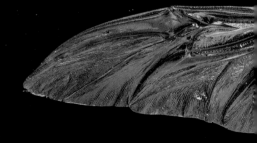

Poplar hawkmoth

Laothoe populi

At rest, the vivid hind wings of the poplar hawkmoth are concealed behind the fore wings. When disturbed, the moth flashes its hind wing to alarm a predator. However, it is very difficult to spot a poplar hawkmoth at rest, with its camouflage colours and its abdomen curled up, increasing its resemblance to withered leaves.

Distribution
Europe

Size
65–90 mm (2½ –3½ in) wingspan

Peanut lantern bug

Fulgora laternaria

Peanut lantern bugs have a large protuberance on the head, in the shape of a peanut with false eyes to resemble that of a lizard or snake. The insect was originally and mistakenly believed to be luminescent (hence its scientific name). When attacked, it protects itself by displaying large, yellow eye spots on its hind wings to frighten away predators and releases a foul-smelling substance.

Distribution
Central and South America
Size
85–90 mm (3¼–3½ in) long

Darwin's beetle

Chiasognathus grantii

Charles Darwin collected this species in Chile during his second voyage and noted that the jaws of the male were not strong enough to produce any pain to his finger, despite being powerful and efficient in combat with other males. The adults feed mainly on the sugary juices from trees. This beetle is uncommon in its natural environment in the southern Andes and is therefore vulnerable.

Distribution
Argentina and Chile

Size
Male up to 90 mm (3½ in), female up to 37 mm (1½ in) long

Imperial jezebel

Delias harpalyce

The upperparts of this butterfly
look much like those of many other
whites (in the family Pieridae): white
with a dark border. The underside,
though, is richly splattered with
colours. The black caterpillars
are clothed with white hairs and
live gregariously in a silk web for
protection. They feed on mistletoes
in eucalyptus forests.

Distribution
Australia
Size
60–70 mm (2½–2¾ in) wingspan

Darkling beetle

Nesioticus flavopictus

This species occurs under the loose rotten bark of dead trees where it feeds on the mycelia of polypore fungi. Its pale markings vary in length and shape and is one reason why scientists use a wide range of external and internal features – body proportions, sculpture of surfaces, shape of eyes, venation of wings and peculiarities of legs – to identify and describe species. Using a single specific feature only (such as body colour) would lead to incorrect classification.

Distribution
Western and Central Africa
Size
15–25 mm (¾–1 in) long

Wax tailed bug

Alaruasa violacea

The wax tailed bug is a fulgorid bug
(family Fulgoridae). Some species
of fulgorid nymphs produce waxy
secretions from special glands on
the abdomen and other parts of
the body. Adult females of many
species also produce wax, which
may be used to protect eggs. Waxy
projections from the abdomen are
especially developed in this beautiful
species from the rainforests of
Central America. Adults and nymphs
feed on tree sap.

Distribution
Mexico and Central America
Size
85 mm (3¼ in) long

Picasso moth

Baorisa hieroglyphica

The species name *hieroglyphica* refers to the striking geometric lines and shapes on the fore wings. Perhaps the shapes resemble a red insect head with antennae and legs, directing a bird's bill towards the wingtips? Or a spider on its web? Although sometimes referred to as the Picasso moth, you might think Miro moth is more appropriate.

Distribution
Northern India, Southeast Asia
Size
50 mm (2 in) wingspan

Cape stag beetle

Colophon primosi

This flightless crawler is from high mountainous areas of southern and western Cape Province. All *Colophon* species are strongly localized in their distribution and usually only occupy one to a few mountain tops. *Colophon* are threatened by extinction due to habitat loss and over-collecting in the past, but the species are now affected by climate change since fires have become more frequent in their habitat due to global warming.

Distribution
South Africa
Size
18–39 mm (¾–1½ in) long

Giant scoliid wasp

Megascolia procer

One of the largest wasps known, *Megascolia procer* attacks and lays its egg on the larva of the huge atlas beetle, *Chalcosoma atlas* and three-horned rhinoceros beetle, *Chalcosoma moellenkampi*. Scoliid wasps have strengthened wing membranes, with the outer quarter finely corrugated, to give rigidity. Light hitting the wings meets interference from a thin layer of chitin and black pigment, giving an intense blue-green iridescence.

Distribution
Java and Sumatra
Size
35–70 mm (1½–2¾ in) body length

Great pied hoverfly

Volucella pellucens

This species, which resembles a bumblebee in size and shape, occurs in most woodland habitats in Europe and Asia. Mimicry of bees in shape and markings is thought to protect them from falling prey to predators, which avoid feeding on true bees owing to their stings. In certain lights the lower part of the abdomen is transparent. The larvae occur inside the underground nests of common wasps.

Distribution
Europe and Asia

Size
15–18 mm (approximately ¾ in) long

Thistledown velvet ant

Dasymutilla gloriosa

Another name for velvet ants is 'cow killers', because of their fearsome sting. Whilst many species advertise their painfulness with bright colours, *Dasymutilla gloriosa* has evolved camouflage to protect it as it roams the bare ground in the desert. It has an amazing resemblance to the fluffy seeds of the creosote bush, a plant found in the same places.

Distribution
Deserts of the southern USA and northern Mexico

Size
13–16 mm (½–¾ in) body length

Tiger moth

Composia credula

Various tiger moths have evolved spotty colour patterns, but no other species are quite as polka-dotted as *Composia credula*. This species was described early on in the history of European exploration of the New World, in 1775, by the illustrious early entomologist, Johan Fabricius. This tiger moth has a close relative in Florida, where it is known as the faithful beauty, or uncle Sam.

Distribution
South America, Antilles

Size
48–64 mm (2–2½ in) wingspan

Temminck's straight-snouted weevil

Eutrachelus temmincki

This is the longest representative of a worldwide species-rich family of straight-snouted weevils. The head of this unusual species looks like it has been stretched, ending in a narrow awl-like 'nose' in the female and widened into a short shovel in the male. The beetles aggregate on large fallen trees, where their larvae dwell. The largest and strongest males have a better chance of mating and leaving offspring.

Distribution
Indo-Malaya

Size
55–80 mm
(2¼–3¼ in) long

Namib darkling beetle

Onymacris sp.

White colouration is unusual among beetles in general and in the large worldwide family of darkling beetles in particular. This remarkable phenomenon is known only in a few species from the Namib Desert. The whiteness is not a pigment but reflection involving microscopic 'bubbles' within the surface of the beetle's hardened fore wings. It may prevent this diurnal species from 'boiling' when ground temperatures reach as high as 45°C (113°F).

Distribution
Namibia

Size
20–32 mm (1–1¼ in) long

Velvet ant

Leucospilomutilla cerbera

This insect gets its species name from Cerberus, the fearsome guard of the gates to the underworld in Greek mythology. Velvet ants are a group of solitary wasps (not ants!) in which the females are wingless, and they have a reputation for their painful sting. The message of fearsomeness is reinforced by the colour pattern, which says 'stay away', and may be mimicking some spider colour patterns.

Distribution
South America

Size
10–10.5 mm (approximately ½ in) body length

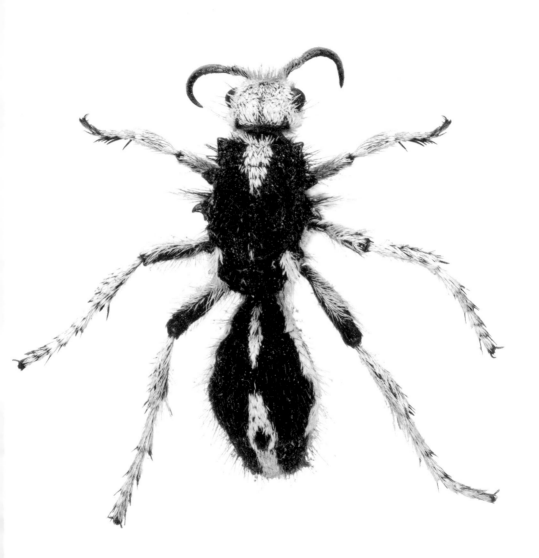

Handsome fungus beetle

Cacodaemon satanas

I am small and vulnerable but well prepared! As its name suggests, the handsome fungus beetle feeds on and develops in various types of tree fungi called polypores, which grow on decaying wood. Some species, particularly from west Malaysia and Borneo, develop sharp spikes over their body for protection from toads, lizards and insectivorous birds. If you were an insect-loving predator, would you risk swallowing this prickly 'hedgehog'?

Distribution
Borneo
Size
12–20 mm (½–1 in) long

The Museum

London's Natural History Museum is not only a tourist attraction, but it is also a world class research institution which employs over 300 scientists and houses many of the world's most important taxonomic collections. Gathered over 300 years and containing more than 34 million specimens, the Museum's vast entomology collection is the oldest and most extensive of its kind anywhere in the world. These specimens are key to telling the history of collecting, the science of taxonomy and the human desire to understand the natural world.

The authors

The authors, all scientists at the Natural History Museum, London, are Gavin Broad, Principal Curator of Insects; Blanca Huertas, Senior Curator of Lepidoptera; Ashley Kirk-Spriggs, Senior Curator of Diptera; and Dmitry Telnov, Curator of Coleoptera.

Acknowledgments

David Lees checked the names of Lepidoptera and supplied interesting nuggets of information, and Judith Marshall and Mick Webb kindly checked the names and text for the small orders sections and Benjamin Price allowed access to the small orders collections of the Natural History Museum, London.